GERMAN ARTIL
1864-1910

Sonja Wetzig

translated from the German by David Johnston

150-mm heavy howitzer of the foot artillery on maneuvers.

Schiffer Military/Aviation History
Atglen, PA

ACKNOWLEDGEMENTS

The author wishes to thank: Mrs. Gisela Reimer, Mr. Walter Reimer and Mr. Wolfgang Fleischer for their support in the research that went into this book. Thanks are also due to Mr. Thiede for his work on the photographs.

SOURCES

Benary. Ehrenbuch der Deutschen Feldartillerie
Bleyhoeffer. Die schwere Artillerie des Feldes
Frobenius. Geschichte des Preußischen Ingenieur und Pionierkorps von der Mitte des 19. Jh. bis zum Jahre 1886
Jany. Geschichte der Königlich Preußischen Armee, Bd. IV
Scheibert, J.   Illustriertes Deutsches Militärlexikon, 1897
Schoen. Geschichte des Deutschen Feuerwerkerwesens...
Schuster, Francke. Geschichte der Sächsischen Armee
Chart, postcard and photo collection of the Bundeswehr Military-Historical Museum Dresden.

PHOTO CREDITS
Fleischer (3)
Military-Historical Museum Dresden (68)
Thiede (9)
Wetzig (6)

TITLE ILLUSTRATION
Bavarian Artillery in the Franco-Prussian War 1870-71.

Contemporary military postcard, a satirical representation of the Heavy Field Howitzer 02. The caption reads: "Soon our exercise will be over, then I'll fly home like a bomb (sic). Greetings from the Wahn Firing Range."

Translated from the German by David Johnston.

This book was originally published under the title, *Waffen Arsenal-Deutsche Feldartillerie 1864-1910* by Podzun-Pallas Verlag

We are interested in hearing from authors with book ideas on related topics.

Published by Schiffer Publishing Ltd.
77 Lower Valley Road
Atglen, PA 19310
Phone: (610) 593-1777
FAX: (610) 593-2002
Please write for a free catalog.
This book may be purchased from the publisher.
Please include $2.95 postage.
Try your bookstore first.

Aerial photo of the Royal Saxon Arsenal and Clothing Office. The arsenal was part of Albertstadt in northern Dresden, one of the largest self-contained barracks installations in Europe. It was built in the 19th Century.

INTRODUCTION

Included in the chart collection of the Military Historical Museum in Dresden is a collection of approximately 400 individual sheets with design drawings and photographic illustrations of guns and vehicles of the Royal Saxon Army in the second half of the 19th Century. The individual sheets were used by artillery workshops and depots as references for the repair and adaptation of equipment already in service. As well they were used as illustrative tables for training. According to the army regulation sheet dated 30 March 1905, the tables could be ordered by any military installation. The tables contained handling notes from various sources, for example instructions from the War Ministry concerning modifications to equipment, or internal regulations for the staff of the artillery depots.

The tables bear obvious signs of decades of use. Their whereabouts until after the Second World War are unclear. The plans probably belonged to the cultural assets which were returned from the Soviet Union in the 1950s. As successor to the Royal Saxon Army Museum, the museum in Dresden received a part of the Soviet war booty.

The chart collection lay sadly neglected in the museum's magazine for twenty years. Although it is incomplete, the exhibit provides a revealing look back at the equipment used by the Saxon Army: guns, carriages, limbers, munitions and rations wagons and other special vehicles as well as ammunition, accessories and the packaging of ammunition. The material presented here is applicable to all contingents of the German Army. As is generally known, the Saxon Army was reorganized on the Prussian model in the course of the unification of the German Reich, which was completed in 1871. Research into the armament of the German Army at this time is a difficult proposition because the German Army

archive was almost completely destroyed in an Anglo-American air raid on Potsdam in 1945. The Saxon sources from the pre-First World War era are thus especially significant in depicting the equipment and armament used by the German Army at that time.

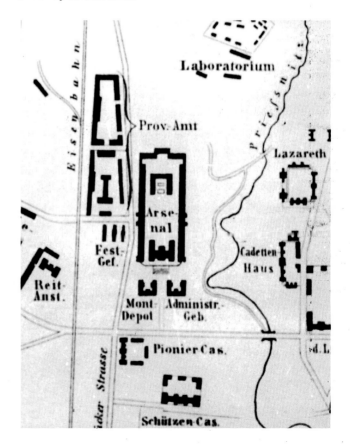

3

ARTILLERY
IN THE SAXON ARMY 1867-1910

With the reorganization of the Saxon Army, which on 7 January 1867 joined the army of the North German League as the XII Army Corps, efforts were made to achieve standardization of equipment for all contingents based on the Prussian example. The result was a number of structural changes within the artillery. General Podbielski became the Inspector-General of Artillery in 1872. All sixteen Saxon foot- and horse-artillery batteries were combined in Field Artillery Regiment 12. The associated organizational separation of the fortress artillery did not follow until 1874.

The army created the 1st Royal Saxon Field Artillery Regiments No. 12 and No. 28. They were also designated as corps and division artillery. To the foot artillery belonged the Foot Artillery Regiment No. 12; subordinated to the artillery brigade were Pioneer Battalion No. 12 and Train Battalion No. 12. In the coming years the mounted field batteries receded further into the background, but they retained some significance as the cavalry's highly-mobile artillery. In the course of time the Royal Saxon Foot Artillery Regiment No. 12 was strengthened by incorporating the crews of the siege and fortress guns. The reorganization of the Saxon Army also saw a gradual reequipping of the artillery. Rifled breech-loading guns had demonstrated their technical superiority over muzzle-loaders in the campaign against Austria in 1866. The experiences from the war of 1870-71 against France showed the need for a general reequipping of the German Army. In 1873 the units were given 88-mm and 78.5-mm field cannon for field evaluation. In 1874 these guns entered service with all German troop contingents as the 90-mm C/73 heavy field cannon with the horse-drawn units and the 80-mm C/73 light field cannon with the mounted batteries. New and innovative features were the cylindrical wedge-type breechblock of cast steel and the gas-tight base, the so-called Broadwell ring (named after its inventor). A range of 6,000 meters far exceeded the military's request for a gun with an effective firing range of 2,500 meters. This increase in firing range was made necessary by the introduction of the Chassepot rifle during the Franco-Prussian War.

The guns were provided with double-wall ammunition. Cartridge-type shells were introduced in 1876, as well as shrapnel and canister ammunition. 1874 saw the introduction of a new gun carriage with sides of pressed cast steel sheet.

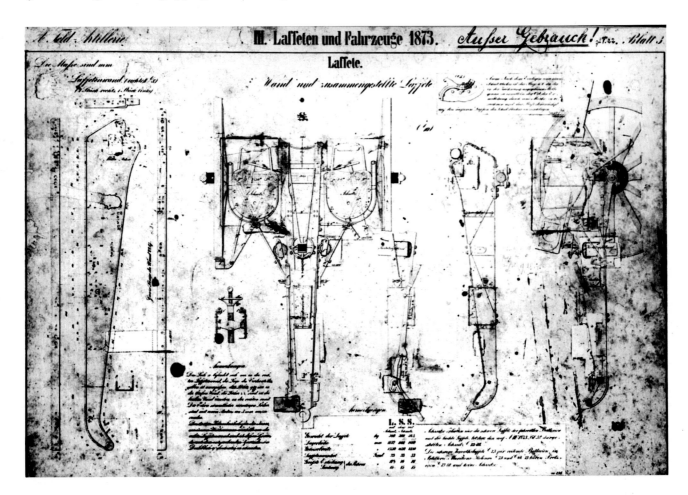

Chart from the collection of the military-historical museum of the German Armed Forces. The hand-written notes are interesting. Depicted is the gun carriage of the C/73. The damage which resulted from improper storage is evident.

There was also much development in the field of siege and fortress artillery. They were equipped with 100-mm and 120-mm cannon, which were later joined by howitzers and the short-barreled 210-mm siege gun/howitzer, as well as 50-mm assault defense guns and 37-mm revolver cannon. 90- and 150-mm siege guns were also used for a short time.

Improvements in gun design, together with new types of gunpowder and explosives, brought a decisive improvement in the effectiveness of the artillery. Firing range and the penetrative ability and explosive power of projectiles were all significantly increased.

In test firings conducted between 1884 and 1887 against different fortress targets, shells filled with trinitrophenol (Shell Filler 88) exhibited a destructive power exceeding anything seen before.

These new qualitative features of the artillery weapons and their growing number, which was reflected in the rap-idly-accelerating formation of new artillery battalions and regiments, bestowed upon this branch of the service an increasingly dominant role in the conduct of land warfare. This applied equally to the field and foot artillery. There were further changes for the artillery at the turn of the century. Improved explosives and new gun and fuse designs allowed the caliber of field cannon to be reduced from 90 to 77 millimeters.

In the German field artillery, the Field Cannon 96, which still lacked barrel recoil when introduced, was the dominant weapon for only a brief time. It was quickly replaced by the 77-mm Field Cannon 96 New Version, which unlike the Field Cannon 96 was equipped with barrel recoil and a gun shield.

From then on modern guns equipped with recoil dominated in the foot artillery: the 150-mm Howitzer 02, the 100-mm Cannon 04 and the 210-mm Howitzer 1911.

Ein Hindernis

This photo illustrates the high degree of mobility possessed by the field artillery. Here a 77-mm Field Cannon 96 (old version) is moved forward into firing position at full gallop. The photo dates from 1903-04.

FIELD VEHICLES

The Military-Historical Museum's Saxon source material described in the introduction also allowed interesting conclusions to be made about the wide variety of field vehicles used by the German Army. In Germany field vehicles were divided into limbers, wagons (with four wheels) and carts (with two wheels). Vehicles were classified as well from the material used in their construction — steel, iron and wooden vehicles — and according to their purpose — ammunition, shell, supply, rations, fodder, baggage or general-purpose wagons. A wartime-strength battery of the field artillery, for example, had six gun and ammunition wagons. As well there were special wagons, for example for transporting supplies, baggage and rations, as well as forge and fodder wagons and the field kitchen. The large variety of vehicles was also typical of the foot artillery. A howitzer battery of four guns consisted of an observation wagon, two gun platoons, two wagon platoons and two ammunition wagons each. The munitions echelon had another four ammunition wagons. The battery also included the combat train, which included the off and spare horses as well as the field kitchen with a supply wagon.

77-mm Field Cannon (new version) in firing position. In contrast to the earlier model, this cannon has barrel recoil and a gun shield. Photo taken during maneuvers in the period before the First World War.

Long-barreled 150-mm cannon (built-up barrel) in firing position. Soon after the outbreak of war older guns such as this were taken from the siege trains in order to compensate for the shortage of artillery at the front. They were used primarily by the reserve artillery.

The large train, which was also part of the battery, included one or two six-horse supply wagons, a two-horse rations wagon, a four-horse fodder wagon, a six-horse field forge and two four-horse spare carriages. On the march such a column was of considerable length. The composition of a heavy howitzer battery was even more complex. In addition, to supply the foot artillery with ammunition there were also light ammunition trains with 24 ammunition wagons and ammunition battalions with 136 ammunition wagons.

The numerous ammunition wagons, supply wagons, fodder wagons and special vehicles, as well as field forges and ambulance wagons, all drawn by two- or four-horse teams, served to bestow the utmost in mobility upon the battalions and thus create tactical advantages. It was a development that was further reinforced by the addition of more specialized wagons in the period before the First World War.

The higher rate of fire of infantry and artillery weapons placed new demands upon supply. Increased firing ranges made necessary much additional equipment (gunsights etc.), in order to achieve the desired precision and effect. Finally, the increased effectiveness of the weapons forced planners to calculate on a greater number of wounded in case of war, casualties who had to be transported out of the combat zone and cared for.

In general it may be said that the numerical strength of military units increased. Providing them with the required supplies was an important task. The numerous innovations introduced into service by the turn of the century, especially in the artillery — special equipment, tools, targeting and aiming devices, and ammunition packaging — can only be hinted at within the scope of this volume. Of course even then the procurement offices were restricted by the need to economize wherever possible. This was evident even in the artillery. There, for example, items of equipment were frequently modified to fit new guns or newly-introduced ammunition.

A horse-drawn howitzer battery of Foot Artillery Regiment No. 19 photographed as the unit assembled for action. The 150-mm Howitzer 02 was the foot artillery's first gun with barrel recoil. It weighed 2,035 kg and had a firing range of 6,050 meters. It is interesting that this battalion of Foot Artillery Regiment 19 was obviously in the process of reequipping when this photo was taken: the 3rd and 4th guns (from the right) are older-model howitzers without barrel recoil.

I. General-Ansichten. 1873.

Schwere Feldkanone C/73.

Truppenübungsplatz Königsbrück (Neues Lager).
Wache am Haupteingang.

Two Type C/73 Heavy Field Cannon in front of the main guardhouse at Königsbrück. The C/73 weighed 1,200 kg. This weapon, which had a caliber of 88 mm, fired shells weighing 7.5 kg equipped with impact fuses. Maximum firing range was 6,500 meters.

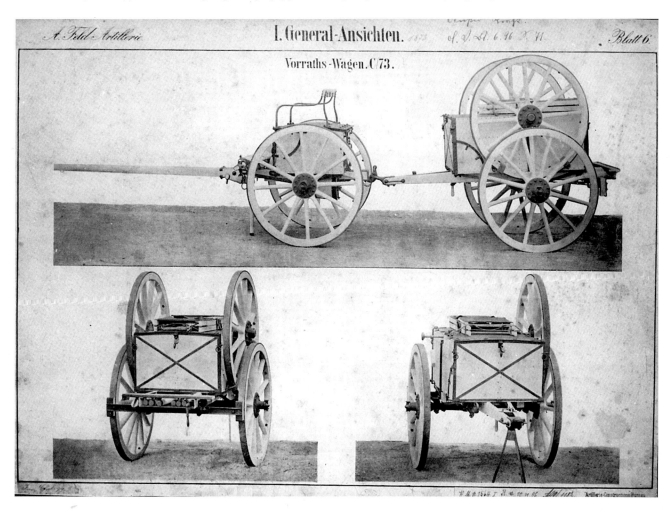

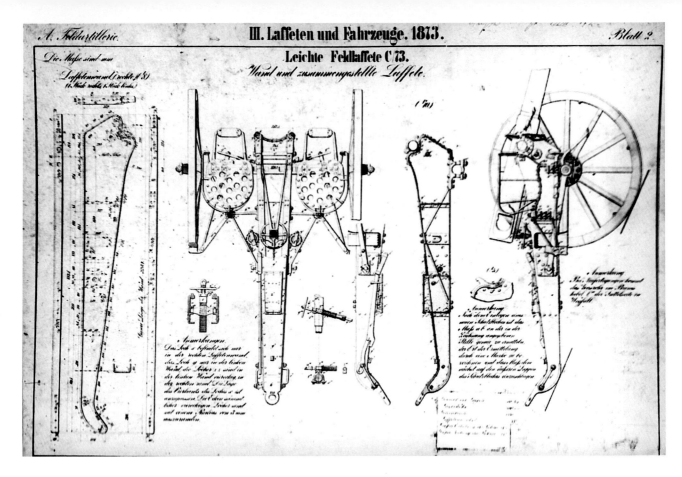

Design drawings for the C/73 Light Field Cannon. This weapon was used by the cavalry's horse-artillery batteries. Firing range was 6,000 meters, muzzle velocity was 465 meters per second.

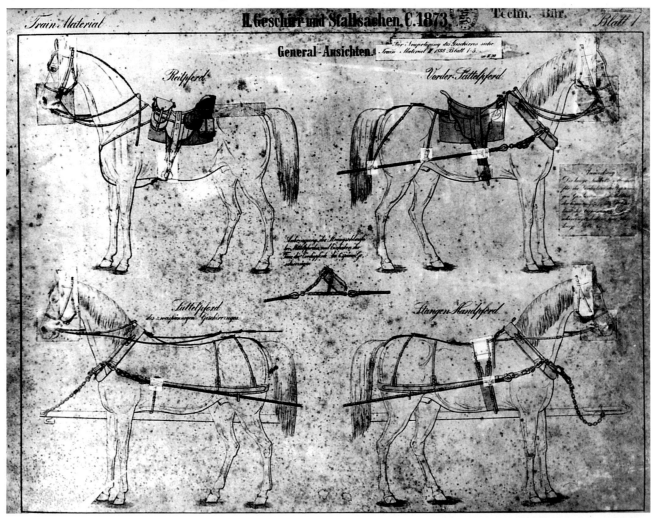

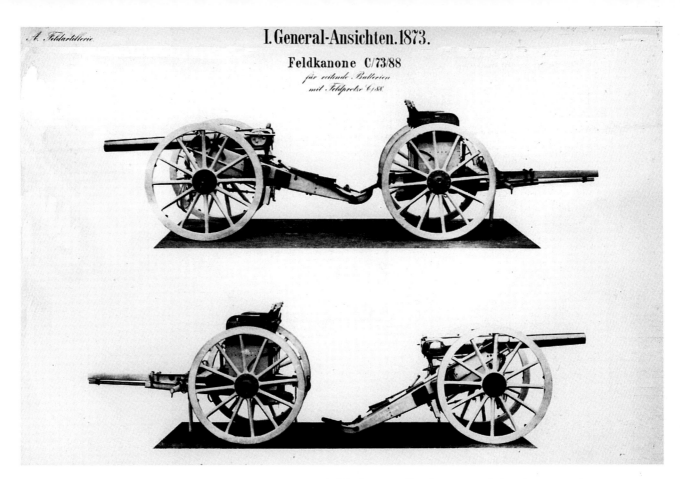

I. General-Ansichten. 1873.

Feldkanone C/73/88
für reitende Batterien
mit Feldprotze C/88

The introduction of the modernized C/73/88 Field Cannon provided the horse-artillery batteries with their first gun in the 90-mm class. The gun weighed 1,055 kg as used by horse-drawn batteries and 980 kg as used by the mounted horse-artillery batteries. The barrel and breech mechanism were the same as the Model C/73.

I. General-Ansichten. 1873.

Feldlaffete C/73/88.
für reitende Batterien
mit Feldprotze C/88

Feldkanone C/73/91

The C/73/91, a further improved version of the 90-mm field cannon, reached the troops at the beginning of the 1890s. The field cannon weighed 1,045 kg, muzzle velocity was 442 m/sec. The gun fired shells weighing 7.5 kg and maximum firing range was 6,500 meters.

A 90-mm C/73/91 Field Cannon photographed during maneuvers. The photo was taken after 1910. Note the field-gray helmet covers already being worn!

Typical scene from a pre-war maneuver. The horse-drawn battery of an artillery brigade, equipped with C/73/91 Heavy Field Cannon, passes infantry moving up to position.

Artillery firing position from the pre-war period. The guns sit side by side in unprotected formation.

The barrel of a C/73/91 Heavy Field Cannon from the collection of the Bundeswehr's Military-Historical Museum in Dresden. It was removed from service on account of a deformation in the forward third of the barrel and assigned to the former Saxon Army Museum. The built-up barrel was made of nickel-steel; the entire barrel with breech mechanism weighed 442 kg.

Artillery equipment park at the Zeithan Troop Training Grounds.

Übersicht

der Entwickelung der Königl. Sächs. Feldartillerie vom 1. Januar 1821 bis 1. Oktober 1901.

Gliederung einer Feldkanonenbatterie (um 1895)

Organization of a horse-drawn battery of the field artillery.

15

The 77-mm Field Cannon 96 entered service with the field artillery shortly before the turn of the century. It was used in batteries of six guns and three ammunition wagons. Caliber was 77 mm, barrel length was 27.3 calibers. The gun fired high-explosive shells weighing 6.85 kilograms a maximum distance of 8,000 meters.

Below: One of the few surviving examples of the Field Cannon 96 is found in the Bundeswehr's Military-Historical Museum in Dresden. The gun carriage was made of cast sheet steel and the barrel — which had 32 grooves — of nickel-steel.

The Field Cannon 96 in the Military-Historical Museum has the barrel number 129; it was built by the Friedrich Krupp AG in 1897.

Ammunition used by the German field artillery until 1896. From right to left: high-explosive shell with impact fuse for the Field Cannon C/64, high-explosive shell with impact fuse for the C/73/88 Field Cannon, field shell with double fuse for the 77-mm Field Cannon, canister round (cutaway model) for the 77-mm Field Cannon 96.

Beim Geschützreinigen.

Kaserne des K. S. 1. Feldartillerie-Regiments No. 12.

Dresden.

1556 Brück & Sohn, Meissen.

Gunners cleaning the Field Cannon 96.

Übergang der reitenden Abteilung d. 1. Feldartillerie-Regiments No. 12 (Königsbrück) über die Schiffsbrücke bei Soebrigen am 20. 9. 05. Sturz eines Pferdes.

4003 Kunstverlag Max Köhler, Dresden-N.

Two-piece ammunition wagons of a horse-artillery battery.

A 77-mm Field Cannon (new version). Photographs from a maneuver during the pre-war period. Introduction of the 77-mm Field Cannon 96 (new version) had to be accelerated because the French artillery had received the 75-mm Model 97 cannon, a modern weapon with barrel recoil which was clearly superior to the early version of the Field Cannon 96.

The field artillery introduced observation wagons in an effort to achieve better firing results.

Schildstütze durch
Bindestrang befestigt

vordere Wagenstütze

hintere Wagenstütze

A gun in difficult terrain. Snapshot from a pre-war maneuver.

Schwierige Auffahrt.

Introduction of the 105-mm Light Field Howitzer produced a decisive improvement in the firepower and operational potential of the German Army's field artillery. Pictured above is the Model 98 without barrel recoil, in the photo below the Model 98/09 with barrel recoil.

THE UNIFORMS
OF THE ROYAL SAXON FIELD AND HORSE ARTILLERY

Koenigreich Sachsen — Feld und Fuss Artillerie

1.   **Field Artillery Regiment No. 12 (Horse Battalion) First Lieutenant in dress uniform**

2.   **Field Artillery Regiment No. 12 (Horse Battalion) Bugler in parade uniform**

3.   **Field Artillery Regiment No. 12 (1st and 2nd Battalions) Gunner in drill uniform**

4.   **Field Artillery Regiment No. 28 (1st, 2nd and 3rd Battalions) Driver in service uniform**

5.   **Foot Artillery Regiment No. 12 Lieutenant**

6.   **Ordnance Lieutenant in dress uniform**

7.   **Ordnance Lieutenant in service uniform**

8.   **Foot Artillery Regiment No. 12, standard uniform, equipped for field march**

GUNS OF THE SIEGE AND FORTRESS ARTILLERY

The 37-mm revolver cannon was introduced in 1884 for close-range defense of fortresses. The Saxon Army produced a unique version of the gun for use in the fortress of Königstein mounted on a modified carriage for the 90-mm cannon.

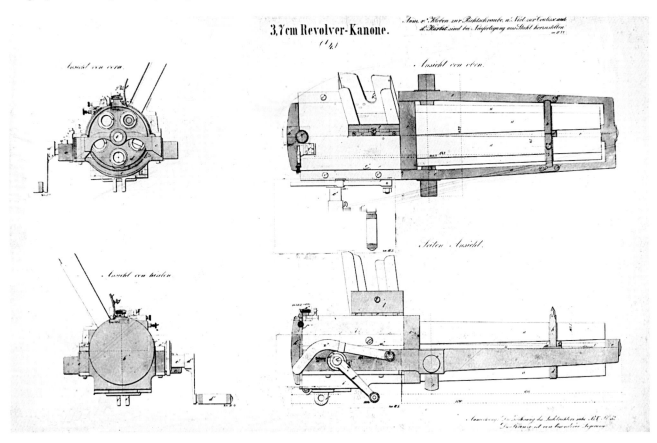

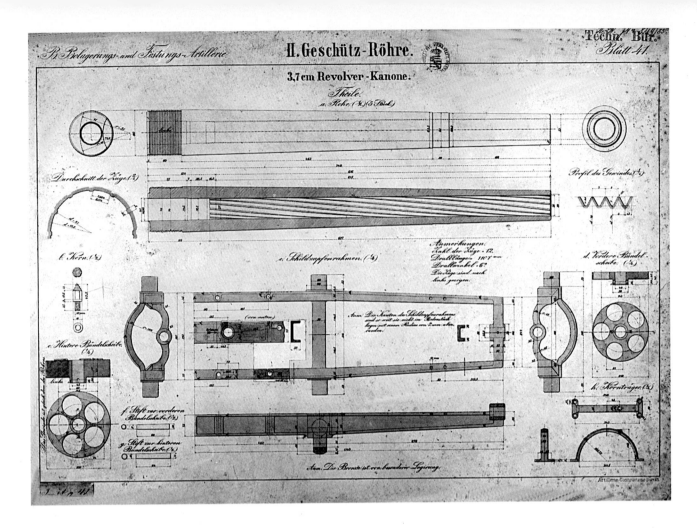

The 37-mm revolver cannon had a caliber of 37 mm and the barrel cluster consisted of five steel gun barrels (weight 211 kg). The weapon fired high-explosive shells with a muzzle velocity of 400 m/sec and a firing range of up to 4,000 meters. It could also fire canister ammunition. The gun was saw use in the First World War as a trench cannon and anti-aircraft gun; later surviving barrels were used to produce the first anti-tank guns.

Beginning in 1889, the fortress artillery was issued a 50-mm rapid-firing cannon, also for close-range defense. They were installed in casemate mounts and mobile armored mounts. When horse-drawn the latter were very mobile and it was intended that they be used to form concentrations of fire in defensive installations. Caliber of the weapon was 53 mm; it fired high-explosive, armor-piercing and canister ammunition.

Above and below:
The barrel of a 50-mm rapid-firing cannon which was used to equip an armored vehicle. It reveals several interesting details for the period, including the falling-wedge breechblock.

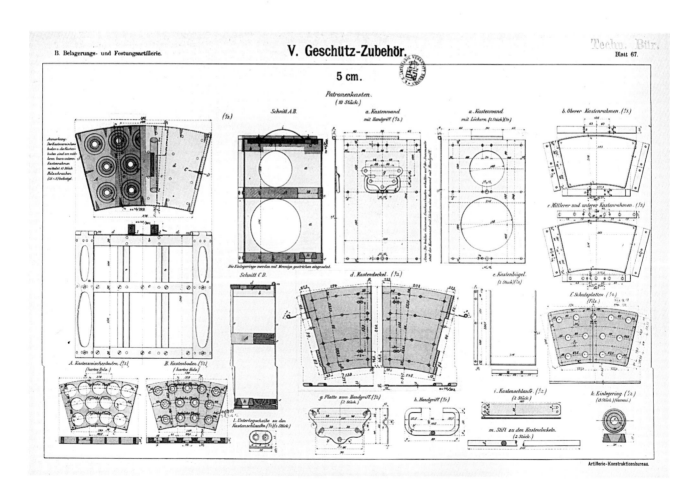

Above and below:
The barrel of the 90-mm Field Cannon was mounted in a special gun carriage for use by the fortress and siege artillery; designated the Heavy 90-mm Carriage, it provided a greater range of elevation.

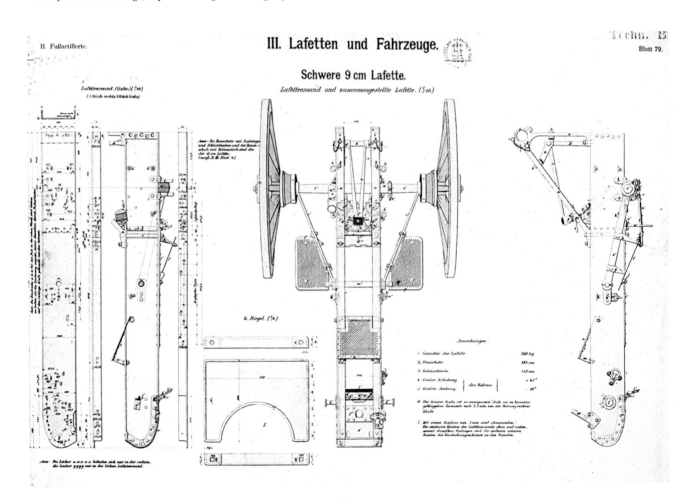

Above and below:
Because of the extreme height differential, use of artillery from the fortress of Königstein in Saxony produced several innovations. In order to be able to engage anticipated targets in the Elbe Valley with direct fire with accuracy, three six-pounder cannon in depression gun carriages were acquired in the mid-1900s. One of the surviving guns belongs to the collection of the Bundeswehr's Military-Historical Museum and may be seen at Fortress Königstein in Saxony, where these photos of the fully-restored gun were taken.

Above and below:
A 90-mm siege gun was felt necessary for engaging fortresses and one was introduced in 1882. It did not remain long in front-line service. Caliber was 88 millimeters, barrel and carriage weighed 196.5 kg. The 7-kg field shells could be fired with four different propellant charges. Muzzle velocity was between 85 and 138 m/sec. Maximum firing range was 1,700 meters.

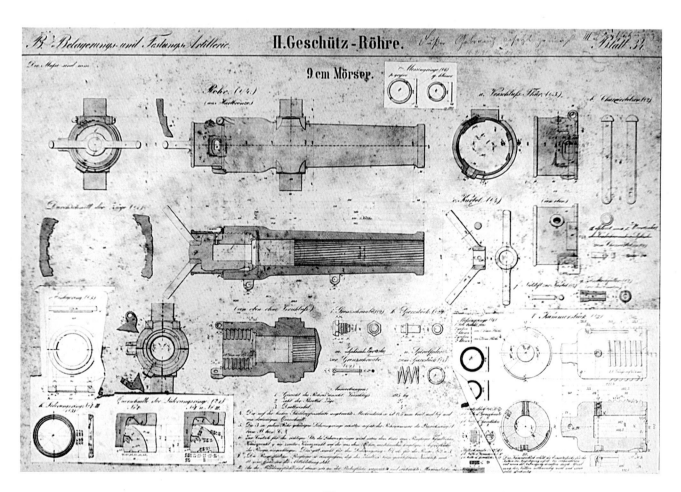

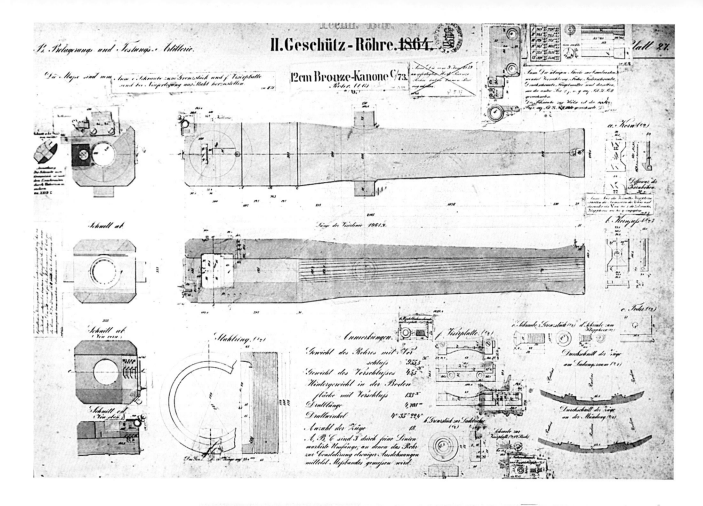

After 1873 the fortress and siege artillery employed the heavy 120-mm siege gun of hardened cast bronze, the 210-mm siege gun, the smooth-bore 150-mm siege gun, and other weapons to demolish ground targets and walled fortifications and for bombardments at ranges of up to 2,000 meters.

Right:
A unique feature of the 150-mm siege gun was that the propellant charge was housed in the interrupted screw-type breechblock. It was otherwise similar to the 90-mm siege gun. Both weapons remained in the inventory for only a short time.

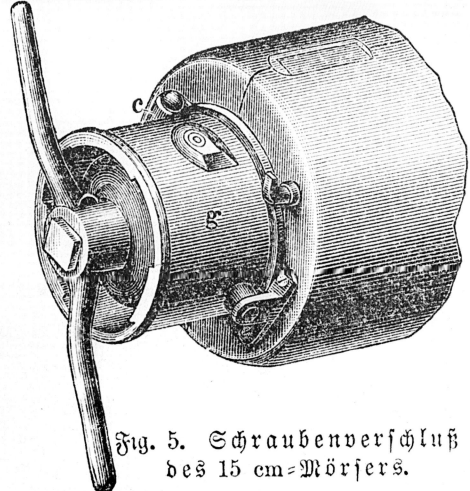

Fig. 5. Schraubenverschluß des 15 cm-Mörsers.

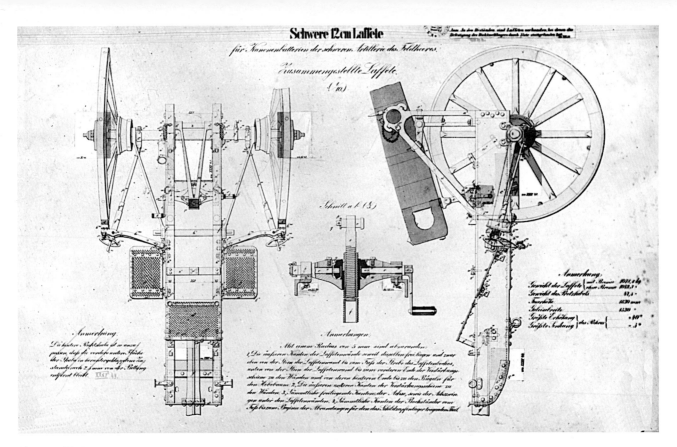

Above and below:
When employed as a siege weapon, the 120-mm cannon was intended to be used against strong earthworks at longer ranges. Maximum firing range was 7,600 meters. Barrel and gun carriage weighed 3,000 kg.

120-mm Turret Howitzer L/13 of 1888-89. Armored guns of this type were built by the Gruson Company of Magdeburg to equip fortresses. In the second half of the Nineteenth Century Gruson was one of the most important manufacturers of gun barrels, conventional and armored gun carriages, and mobile and stationary armored turrets.
Bottom:

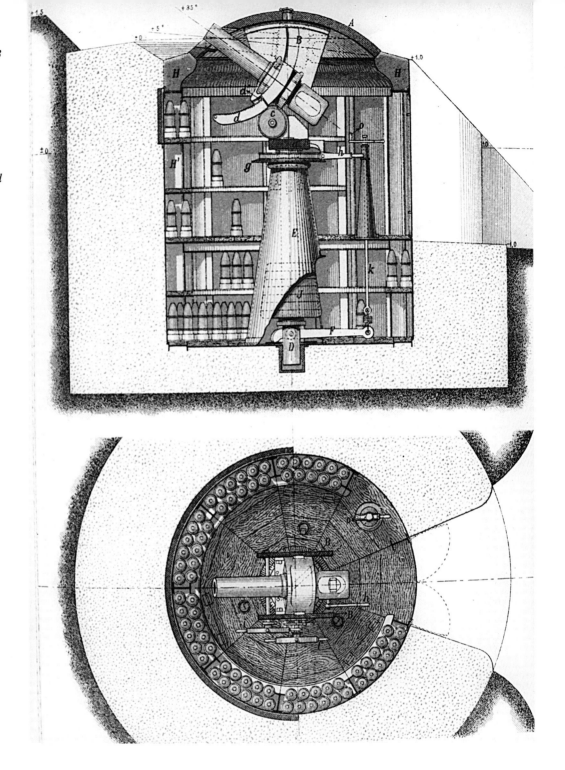

Below:
Barrel cleaner for the 210-mm siege gun.

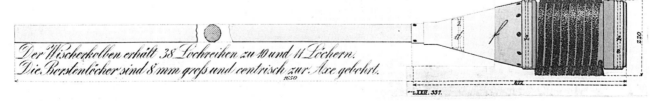

Above and below:
For a long time the 150-mm bronze cannon was the German Army's main gun for delivering indirect fire. Both photographs show the gun in Königstein fortress in Saxony.

Above and right:
The life story of a cannon:
This cannon initially formed part of the armament of Königstein fortress in Saxony, from there it went to Dresden where it was placed in front of the main entrance to the arsenal. After the Second World War it was taken to the Soviet Union as war booty. The 1950s saw it returned to the then German Army Museum in Potsdam. The cannon was then displayed in the GDR's army museum in Dresden and today it is among the best pieces in the artillery collection of the Bundeswehr's Military-Historical Museum. In 1990 it was returned to the fortress of Königstein in Saxony.

150-mm iron cannon with wedge-type breechblock in siege carriage in the display halls of the Saxon Army Museum (pre-1945).

Below:
Lead figure diorama from the display "Fortress Construction and Military History 1806-1945" of the Bundeswehr's Military-Historical Museum in the fortress of Königstein. It depicts the siege of Paris in the war of 1870-1871.

A work detail with four 150-mm cannon outside the walls of the fortress of Königstein.

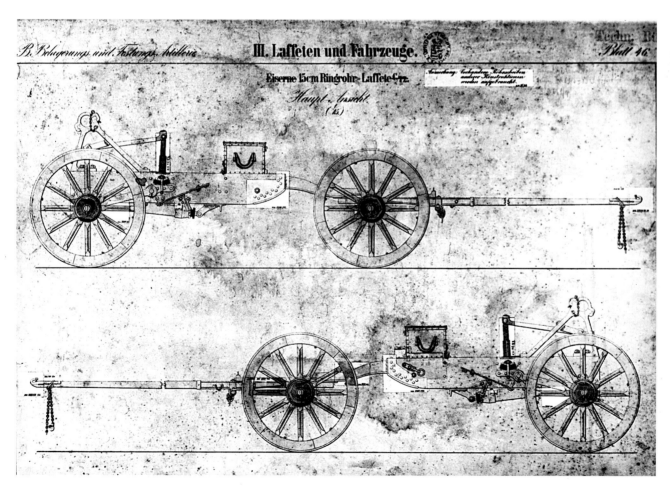

Lange 15 cm Kanone.

Above and below:
Long-barreled 150-mm cannon in a firing position laid out according to the manual. The gun was introduced into service in 1891 and fired shrapnel and long shells (42.3 kg). Maximum firing range with impact-fused shells was 10,000 meters and with powder-train fuses 8,200 meters. Muzzle velocity was 495 m/sec, barrel weight 3,365 kg.

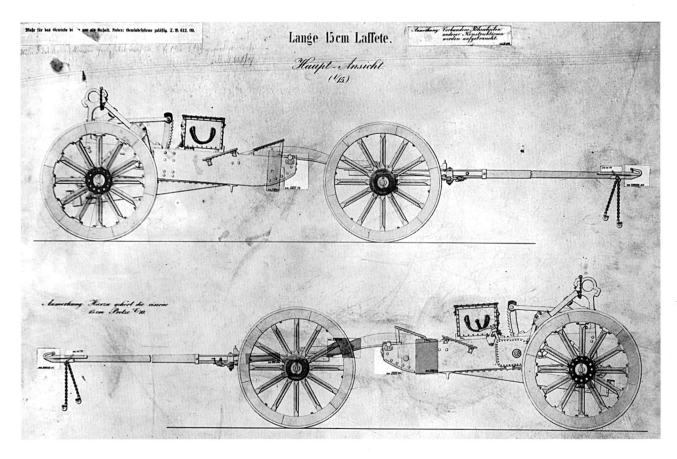

Though obsolescent, long-barreled 150-mm cannon were used in the First World War. Pictured here is one such gun belonging to the 8th Battery of Reserve Foot Artillery Battery No. 20.

Below:
The photo shows 150-mm cannon in the front line, firing over the breastworks. Behind them are 210-mm howitzers for providing indirect fire.

Above and below:
The 150-mm Heavy Field Howitzer entered service with the foot artillery just prior to the turn of the century, initially without barrel recoil. The gun was transported by a team of six horses. Each battery included four heavy field howitzers and eight ammunition wagons.

Above and below:
The 150-mm Heavy Field Howitzer 02, which had barrel recoil, represented a significant improvement. Caliber was 149.7 mm, barrel length 1,770 mm, equivalent to 12 calibers. Weight in firing position was 2,035 kg. The gun fired shells with impact fuses (39.5 kg). The field howitzer had a firing range of 6,050 meters. 840 howitzers were on strength in July 1914.

Schwere Feldhaubitze 02.

Above and below:
The 210-mm bronze howitzer in an iron siege carriage; the barrel was 2,042 mm long and weighed 3,078 kg. Caliber was 209.3 mm.
Firing range was 4,200 meters with a 146.6-kg high-explosive shell (muzzle velocity 290 m/sec) and 4,200 meters with a 79.1-kg shell.

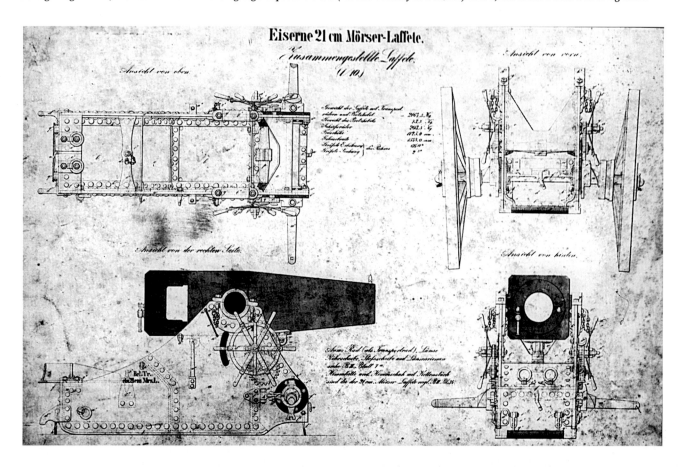

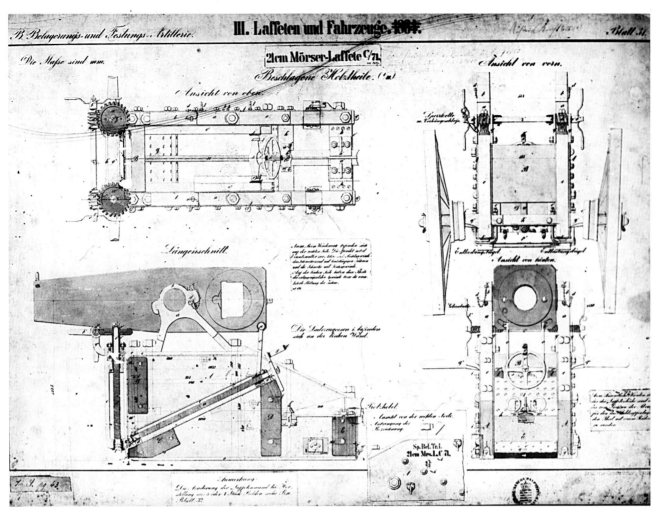

In the beginning wooden carriages were used for 210-mm howitzers,
like the 210-mm Howitzer Carriage C/71.
The wooden carriages proved incapable of handling the stresses that
resulted from increases in firing range and shell weight.

The collection's design drawings also include the great variety of gun
accessories. In the photo: design and dimensions of the shell basket
for the 210-mm howitzer.

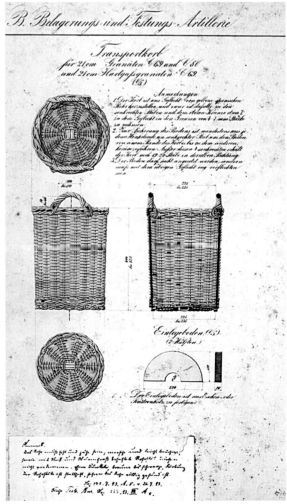

210-mm howitzer in transport position.

210-mm howitzer in firing position.

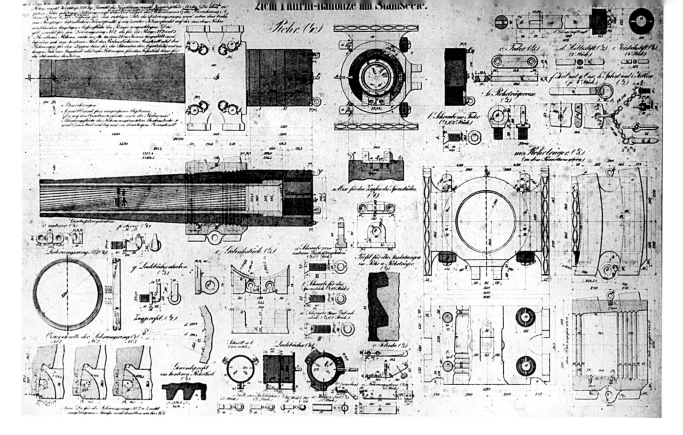

The 210-mm turret howitzer with steel bore was designed to equip fortresses.

The 210-mm Steel Howitzer 99. Caliber was 211 mm, weight in firing position 4,820 kg, shell weight between 83 and 120 kg. This gun saw service in the battles of the First World War.

21 cm Mörser in Feuerstellung.

Like other guns, the 210-mm howitzer was also re-engineered to incorporate barrel recoil and the new version entered service in 1911. Guns of this type formed the backbone of the foot artillery battalions.

Below:
The long-barreled 210-mm howitzer, here in the possession of the Bundeswehr's Military-Historical Museum. The gun was built during the First World War.

A SELECTION OF THE MOST IMPORTANT GUNS
OF THE GERMAN FIELD, SIEGE AND FORTRESS ARTILLERY
BETWEEN 1864 AND 1902

37mm	Revolver cannon in battery mount	Siege and fortress artillery
37mm	Revolver cannon in casemate mount	Siege and fortress artillery
50mm	Rapid-firing cannon in casemate	Siege and fortress mount artillery
50mm	Rapid-firing cannon in retractable	Siege and fortress armored mount artillery
60mm	Rapid-firing cannon	Siege and fortress artillery
77mm	Field Cannon 96	Field artillery, mounted and horse-drawn batteries
77mm	Field Cannon 96 New Version	Field artillery, mounted and horse-drawn batteries
80mm	Light Field Cannon C/64	Field artillery, mounted batteries
80mm	Light Field Cannon C/67	Field artillery, mounted batteries
80mm	Light Field Cannon C/73	Field artillery, mounted and horse-drawn batteries
90mm	Smooth Bore Cannon	Siege and fortress artillery
90mm	Heavy Field Cannon C/73	Field artillery, horse-drawn batteries
90mm	Cannon in Heavy Carriage C/64 Siege Carriage C/64	
90mm	Cannon in Heavy Carriage C/69 Siege Carriage C/69	
90mm	Heavy Field Cannon C/73/88	Field artillery, mounted batteries
90mm	Heavy Field Cannon C/79/91	Field artillery, horse-drawn batteries
90mm	Cannon in Heavy Carriage	Siege and fortress artillery
90mm	Siege Gun in Heavy Carriage	Siege and fortress artillery
100mm	Turret Cannon	Siege and fortress artillery
100mm	Cannon	Siege and fortress artillery
105mm	Light Field Howitzer 98	Field artillery, horse-drawn batteries
120mm	Bronze Cannon C/73	Siege and fortress artillery
120mm	Heavy 120-mm Cannon	Siege and fortress artillery
120mm	Heavy 120-mm Cannon With Steel Bore	Siege and fortress artillery
150mm	Short-Barreled Iron Cannon C/64	Siege and fortress artillery
150mm	Bronze Cannon	Siege and fortress artillery
150mm	Steel Cannon (Short-Barreled 150-mm Cannon)	Siege and fortress artillery
150mm	Cannon (Built-Up Barrel)	Siege and fortress artillery
150mm	Long-Barreled Cannon (Built-Up Barrel)	Siege and fortress artillery
150mm	Heavy Howitzer	(originally field artillery)
150mm	Long-Barreled 150-mm Cannon	Siege and fortress artillery
150mm	Turret Howitzer	Siege and fortress artillery
150mm	Heavy Field Howitzer 02	Siege and fortress artillery
210mm	Short-Barreled Cannon	Siege and fortress artillery
210mm	Turret Howitzer with Steel Bore	Siege and fortress artillery
210mm	Heavy Howitzer C/70	Siege and fortress artillery
210mm	Heavy Howitzer C/70	Siege and fortress artillery
210mm	Bronze Heavy Howitzer	Siege and fortress artillery
210mm	Heavy Howitzer with Steel Bore and Steel Carriage	Siege and fortress artillery
280mm	Cannon (Built-Up Barrel)	Siege and fortress artillery
280mm	Iron Heavy Howitzer	Siege and fortress artillery
280mm	Bronze Heavy Howitzer	Siege and fortress artillery

TYPICAL FIELD VEHICLES OF THE GERMAN ARTILLERY

Vorrathswagen C/69.

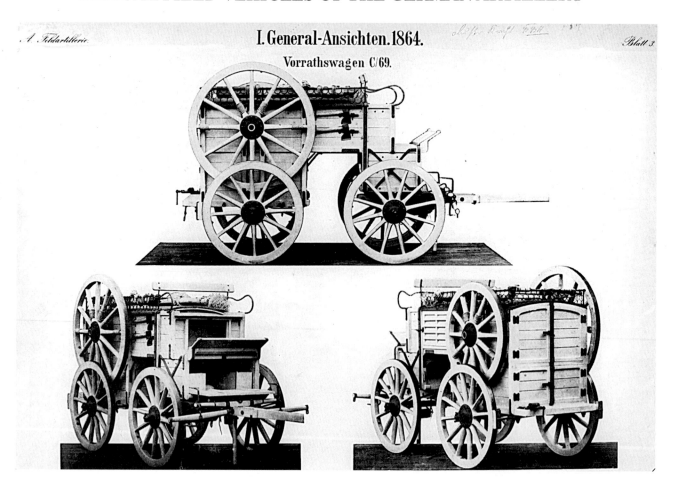

Feldschmiede C/69.

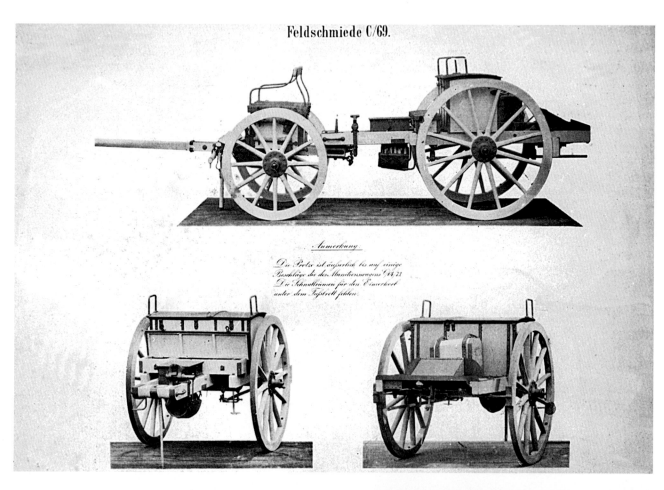

Anmerkung

Die Protze ist äusserlich bis auf einige
Beschläge die des Munitionswagens C/73.
Die Schnallriemen für den Eimerbord
unter dem Fußtritt fehlen.

Vorrathswagen C/88.

Below: Model 95 baggage wagons and ammunition wagons at the troop training grounds at Königsbrück, Saxony.

The Spielberger German Armor & Military Vehicles Series

Four volumes in the classic series by reknowned German panzer historian Walter Spielberger are now available in new English editions. Known for his emphasis on detail, Spielberger chronicles each production variation and later modifications. Line drawings by Hilary Doyle complement the discussion of each model type.

Panther
& Its Variants
WALTER J. SPIELBERGER

Size: 8 1/2" x 11"   288 pages hard cover
over 460 photographs
ISBN: 0-88740-397-2 $39.95

Sturmgeschütz
& Its Variants
WALTER J. SPIELBERGER

Size: 8 1/2" x 11"   256 pages hard cover
over 240 photographs
ISBN: 0-88740-398-0 $39.95

Panzer III
& Its Variants
WALTER J. SPIELBERGER

Size: 8 1/2" x 11"   168 pages hard cover
over 200 photographs
ISBN: 0-88740-448-0 $29.95

Panzer IV
& Its Variants
WALTER J. SPIELBERGER

Size: 8 1/2" x 11"   168 pages hard cover
over 200 photographs
ISBN: 0-88740-515-0 $29.95